SMART FOOD SAFETY IN A GLOBAL REGULATORY ENVIRONMENT

Barbara Rasco

**University of Wyoming,
Laramie, WY 82070**

Table of Contents

Author's Bio   4
SUMMARY   5
THE CURRENT REGULATORY ENVIRONMENT   5
MARKET DRIVEN FOOD SAFETY   9
AGRICULTURAL PRODUCTION AND SMART   9
SMART TRACEABILITY ACROSS THE SUPPLY CHAIN   12
IoT (Internet of Things)   12
AI (Artificial Intelligence)   13
SMART Applications in Food Processing   14
SMART and Novel Food Development and Personalized Nutrition   15
SMART AND GOVERNMENT STRATEGIES AND PROGRAMS   15
The OECD Countries and European Union, Generally   15
Adoption of SMART Food Safety in Several Countries   16

    Australia   18
    Brazil   18
    Canada   19
    China   20
    Columbia   21
    India   22
    Japan   23
    Republic of Korea   23
    New Zealand   24
    Singapore   25
    United Kingdom   25
    United States   25
    European Union   27
    Conclusion   29

ACKNOWLEDGEMENTS   31
References   32

Author's Bio

Barbara Rasco is an attorney specializing in food law. She is an internationally recognized food safety expert with more than forty years' experience in 37 countries assisting companies across the globe comply with US regulatory requirements. She has provided legal and technical assistance to numerous colleagues in an additional 45 countries supporting economic development of agricultural and food businesses.

SUMMARY

Food safety technologies are evolving to incorporate enhanced reliance on artificial intelligence and IoT devices. This will involve sharing large amounts of information along the supply chain to and between regulatory bodies creating technical, business and legal challenges. SMART (self-monitoring, analysis and reporting technology) integrates computing and telecommunication technologies into a range of environments, devices and domains that previously did not have such capabilities, in this case food analysis, food processing monitoring, and supply chain controls. Adopting generative artificial intelligence and enhanced utilization of data analytics could lead to a more effective and simpler management of food safety programs for the private sector and governmental entities as SMART technologies become more prevalent. Many governments continue to focus on integrating technology into regulatory regimes, this is in addition to incorporating leadership development to improve the food safety culture overall. The United States Food and Drug Administration is ushering in a new era of stronger food safety practices as it moves forward with implementation of the Food Safety Modernization Act and much of this is technology based. Korea, Japan, Singapore, Canada, China and India are adopting similar approaches to food regulation and trade. The European Union is emphasizing development and implementation of SMART technologies to address resource scarcity, climate change, environmental sustainability and malnutrition with food safety taking a secondary position in policy. India and Brazil are evaluating how SMART could improve nutrition and food security, promote development of novel foods and make food production more efficient while promoting environmental sustainability. Many companies are adopting SMART to improve safety and efficiency, enhance nutritional quality, reduce waste, and enhance competitiveness. Investment by firms and governments in research and development and technology implementation is spotty. Challenges with adopting SMART include coordination across the supply chain, staying up to date with system security and having skilled individuals among the different stakeholders who can make this coordination possible. As always, the attendant challenges with cost and the interoperability between systems along with the risk of cyberthreats and data breaches remain.

THE CURRENT REGULATORY ENVIRONMENT

Technological advances that could improve food production, sustainability and product safety are key to the vision of food safety agencies. New technologies are revolutionizing agricultural production. Robotics and automated equipment will reduce reliance on manual labor and incorporation of artificial intelligence will revolutionize business processes (Charlebois et al., 2024). Regulatory agencies are beginning to anticipate that farmers can make real-time decisions affecting food safety through their ability to measure and predict environmental factors such as weather patterns, animal movement and water quality that could impact such things as produce contamination and supply chain optimization. Fully automated feeding and climate control systems in animal husbandry are making manual systems obsolete and have the potential to improve animal welfare, reduce disease risk and from this animal drug use through predictive analytics. Food safety agencies in the 38 Organization for Economic Cooperation and Development (OECD) countries and key partner countries are leading this global effort. Adopting smarter food safety in agriculture should improve consumer confidence in the sustainability and safety of the food system (FDA, 2024).

Consumer expectations about food safety and the nutritional quality of their food have changed dramatically as eating patterns and food preferences have changed. With this has come a change in attitude as to who is responsible for the safety of our food evolving as socially liberal policies have been adopted by most capitalist countries (Longley, 2020) causing consumers to believe that governmental regulations and enforcement of regulations is primarily responsible for food safety not the inherent dedication of companies involved in food production to make a safe product. In social liberalism, increasing power is transferred from individuals to a central government to address issues of public health, a trend that has accelerated since the 1960s moving away from a reliance on individuals or society at large to meet public health and food safety goals. In the United States, social liberalism garnered a foothold in the early 1900s in the food safety realm with the passage of two laws, the Pure Food and Drugs Act and the Meat and Poultry Act in 1906 signed by President Theodore Roosevelt on the same day, laws that prohibited the sale of adulterated food with provisions to eliminate unsanitary conditions in food processing facilities. Similar inspection programs for poultry (1957) and eggs (1970) were put into place. Social liberalism grew under the New Deal with the Franklin Delano Roosevelt Administration in the 1930s expanding the role of the federal government in all aspects of life and promulgating protective legislation along with higher taxes to support the central government to finance its regulation of the marketplace including that for food. With food this involved passage of the Food Drug and Cosmetic Act of 1938 extending FDA authority to cosmetics and medical devices, authorizing FDA to conduct inspections, and adding injunctive relieve as a penalty for food regulatory violations in addition to product seizures and criminal prosecution. In 1988, the FDA became an agency under the Department of Health and Human Services led by a Senate approved Commissioner appointed by the President. FDA has the authority to order a food pulled off the market for safety reasons (a food recall), assess and collect fees related to the recall, seize contaminated foods, arrest and criminally prosecute responsible persons in serious cases, conduct an extended audit of company practices, and suspend the registration of a food operation so that it cannot sell food which leads to a temporary and potentially permanent shutdown of the business.

Food safety statutes promulgated in the 2000's expanded FDA authority to address newly identified risks. Following the September 11, 2001 attacks, concerns about the safety of the food supply and food and agricultural infrastructure grew and the Public Health Security and Bioterrorism Preparedness and Response Act (2002) addressed some of these concerns. Allergen labeling for proteins including peanut, soy, milk, eggs, fish, crustacean shellfish, tree nuts followed in 2004, with undeclared allergens now being the primary cause of food recalls in the United States. The Food Safety Modernization Act (2011) greatly expanded the scope of FDA authority through 7 major rules providing new enforcement authority over farms, new requirements for animal foods and fresh produce, expanded regulations for imported foods with required foreign supplier verification programs, audits including 3rd party certifications, new regulations for intentional adulteration or food defense, and new provisions for the sanitary transport of food albeit under the jurisdiction of the Department of Transportation and not the FDA. All provisions of FSMA require traceability, more sophisticated process monitoring and product testing, and a higher level of employee training. Developing regulations to implement FSMA has been a Herculean task and took until 2022 to complete because of the complexity of the statutory provisions and confusion surrounding them. This final rule includes a specific requirement for traceability records for certain high-risk foods.

Social liberalism is front and center in the EU plans for food safety and food systems development across the member states reflecting the continuing push across the European Union to develop a common set of values despite wide geographic disparities (Ulea, 2023) and different cultural values with Northern Europe on a whole holding more progressive views (Berkowitz et al., 2023). To meet the European goal of a unified food safety system, there needs to be leadership at the country and regional level to develop a shared vision for a food safety culture, this continues to evolve. This shared vision will by necessity require there to be agreement on the specific types of technology to be adopted to manage food safety and to address concern about either the safety or sustainability and the extent to which member states will retain sovereignty over decisions affecting their individual food markets. Across the EU there remains disagreement among member states over the use of genetic technologies to modify plant and animal traits; along with the use of agricultural chemicals, food additives and types of packaging that are acceptable in certain member states but not in others. If SMART is to be widely adopted across the EU, an easy to use, stable and secure data platform that is multilingual and allows for the seamless transfer of information between users across the member states will be necessary if a unified SMART food safety system is to be established.

Market integration with the adoption of a single set of unified food safety standards under a Single European Market model across the continent has had a high degree of success but several challenges remain. The establishment of a continent-wide free-trade zone across Europe in the 1990s with the adoption of the euro as a common currency in 1999 led to a generally positive attitude toward market integration among European consumers. This integration shifted responsibilities for food safety from individual nation states prior to 1990 to a Single European Market promoting unified standards forcing integration and harmonization of safety procedures for food production across Europe (Halkier & Holm, 2006). The first major challenge to this single market model was the crisis in confidence caused by bovine spongiform encephalopathy (BSE) in 1989 that was tied seven years later in 1996 to the fatal but extremely rare human disease Creutzfeldt-Jakob Syndrome (CJS). This rocked beef markets globally with repercussions remaining 20 years later. An EU ban on exports of British beef to member states lasted for 10 years (van Zwanenberg & Millstone, 2005). Periodic bans and trade restrictions continue to pop up from BSE around the globe, some legitimate and some politically driven protectionism (Akman et al., 2018). The drop in consumer confidence in the safety of their food following BSE outbreaks resulted in a reorganization of food safety administrations across European countries and at the EU level. This reorganization has gone well in some countries but less so in others, not so much due to the ability to integrate EU requirements into local law, but due to administrative barriers (Halkier & Holm, 2006) and in some cases, the inability of regulatory agencies in certain countries to adopt the technology needed to implement better food safety practices.

Food security is tied to the cost of food and availability of food and cost-effective management of the food supply chain at the state level and the cost of regulatory compliance, including food safety compliance plays an important role in defining whether people in general, or specific individuals within a population will have enough to eat. Food processing costs made up about 25 cents of every dollar a consumer spent on food at the grocery store in 2021 and this ratio has remained constant from 2012-2021 (GAO, 2023) meaning that increases in the cost for food are coming from other factors including input and marketing costs, and impact from climate events and animal and plant diseases such as avian influenza and citrus greening disease. To the extent that SMART can provide predictive models to determine the impact of disease or climate effects on food supply, these technologies will provide an important contribution to reducing negative environmental impacts food security. Similarly, enhanced ability to predict and monitor costs across the supply chain could provide means to get food and food components to where these are needed more efficiently and at a more affordable price could alleviate food security to some extent and reduce waste.

The level of food security in different markets is greatly affected by governmental policies and utilization of predictive modeling to determine how food supplies would be impacted in specific markets would be useful for allocating governmental resources to ensure people have enough to eat. How these models are developed will be determined in part by governmental policy. Food security is driven by governmental policies and tends to fall into one of 3 models (Berkowitz et al., 2023). The first model for food security under a liberal welfare regime such as in the United States is characterized by a free market orientation with a "residualist" ethos where the government only steps in to meet "left over" needs that cannot be addressed by the market, community or family (Berkowitz et al., 2023). Social democratic welfare regimes such as Norway use a "universalist" policy, in the sense that food security policies are meant for everyone regardless of income or occupation, and do not require an individual to wait until their needs are great enough to establish eligibility; programs like this are designed to prevent individuals from become destitute from old age or poor health (Berkowitz et al., 2023). The third model is a "corporatist" regime (*e.g.*, Germany) emphasizing the role of sector stakeholders (*e.g.*, the state, industrial sectors, and unions) and how they must work together to promote social stability and promote family-focused benefits, policy segmentation by occupation, and involvement of non-governmental stakeholders in program design and administration. Entitlement to benefits often stems from a person's work or service to the state (Berkowitz et al., 2023). Berkowitz found that individuals living under corporatist or social democratic regimes had on average, a substantially lower probability of food insecurity than those living under liberal regimes so it is likely that implementing SMART that would affect accurate predictive modeling for food will initially be more successful in countries with a corporatist or social democratic regime.

Clearly changes in the regulatory environment come with a need for both analytical and data handling improvements by governmental agencies if the regulated industry is to comply with updated safety and related market requirements. Fortunately, this has been recognized by many food safety agencies who are seeking the necessary funding to adapt to the demands of the evolving global food market and an expectation that both food safety and food security needs can be met. This is not simply adoption of innovative technologies but requires a change in leadership approach for agencies to work in partnership with industry to change the culture to focus on important and emerging food safety issues and market trends without the risk of enforcement action. As an example, FDA states in its new policy the need for transformative technologies in food science to provide safe and nutritious foods (FDA, 2024). Analytical methods to identify intentional adulteration resulting from fraud, specifically analytical techniques to track credence attributes such as source, GMO free, natural, environmentally friendly are becoming available and are now incorporating features to track carbon footprint and water usage, and the more conventional cultivation conditions such as shade grown and organic are driving SMART food safety for certain high value foods such as specialty oils, honey, coffee and spices. Tracking intentional adulteration across the value chain using end to end traceability, potentially to the item level using blockchain systems and cloud based digital compliance documentation for multiple parties across the supply chain is becoming more common.

MARKET DRIVEN FOOD SAFETY

Further smart food safety is market driven. For example, consumables that transcend medical foods involving development of customized products that cloud the distinction between supplements and drugs with foods meeting the needs for personalized nutrition will require analytical methods that can track not only genetic or biochemical alternations to food ingredients but in addition means to effectively monitor physiological or genetic changes in the people who consume them to insure that these products are wholesome, meet label claims for efficacy and are not fraudulently represented. Plant based proteins plus new foods and ingredients generated in tissue culture or by fermentation such as cell-based meats are the first in a series of novel foods that pose potential food safety challenges but ones that sophisticated process monitoring, traceability and rapid analytical methods for analytes and food additives that traditionally have not been components of interest in foods will be needed. Designing cell-based meats with specified lipid profile or protein composition or using these engineered foods as a delivery system for drugs taking advantage of an individual's unique gastrointestinal physiology to improve absorption (Varum et al., 2013) are exciting areas of exploration. New and developed biochemical engineering processes for creating ingredients such as probiotics, prebiotics, vitamins, flavors, and colorants require SMART rapid analytical methods to ensure quality and safety (FDA, 2023).

Consumer demand for plant-based and alternative meat and dairy protein sources is increasing with alternative products growing in market share over the past ten years due to consumer concern about the environmental impact of animal agriculture and shifts to what are perceived to be healthier foods. Although plant-based alternatives may have a smaller environmental food print at the point of cultivation than an animal protein alternative, these foods are so highly engineered that the environmental impact in terms of agricultural resource utilization and energy and water demand may not be lower. Alternative proteins made by cell culture, from plant or microbial proteins involve sophisticated formulations and bioengineered ingredients. Exciting new developments in bioengineering make it possible to produce proteins from microbial sources with desired food properties. For example, engineered cyanobacteria can produce foreign proteins organized as nanofibers providing meat like textural fibers (Zedler et al., 2023). These foods meet the demand of consumers who are flexitarian, vegetarian or have concerns about animal welfare. Depending on how these proteins are produced and marketed, animal protein alternatives may improve food security in certain communities. Because of the sophistication involved with the production of plant or fermentation-based proteins, SMART technologies will be required so that these alternative protein foods can be made in an efficient and cost- effective manner.

AGRICULTURAL PRODUCTION AND SMART

Agriculture and food production are rapidly adopting SMART technologies and artificial intelligence applications. The most widely used applications at this point involve integrating predictive analytics into crop production systems and into quality assurance and logistics operations. Precision agriculture utilizes machine learning models utilizing output from sensors, satellite imagining and visual output from autonomous vehicles including tractors, harvesters and drones to gather real-time data on soil conditions, crop density and maturity, pest control and to detect early signs of disease, nutrient deficiencies and water stress. This makes it possible for farmers to target interventions such as precise fertilizer and pesticide application and optimize irrigation and harvest timing optimizing yields (Bruun-Jensen et al., 2024). Artificial intelligence is well suited for predictive analytics based on analysis of historical data and the identification of patterns making it possible to develop predictive models to forecast weather conditions, crop yields, and market trends (Bruun-Jensen et al., 2024). Global firms are providing open-source tools for farmers to determine how extreme weather events can affect key crops.

Precision farming or precision agriculture is rapidly evolving and uses sophisticated real time monitoring and data analytics optimizing farming practices to enhance efficiency, minimize waste, and mitigate the environmental impact of various agricultural practices (TraceXTech, 2024). Techniques such as GPS-guided tractors, drones, and chemical sensors collect data on soil conditions, crop health, and weather patterns. Results then inform decisions about the timing and location of applications providing a precise means to optimize use of water, fertilizers, and pesticides. Precision farming can optimize resource use, reduce unnecessary inputs, minimize water use, optimize soil nutrient profile and reduce cost of expensive inputs (Bruun-Jensen et al., 2024). Improving agricultural practices through SMART can reduce the risk of nutrient runoff, soil erosion, and overuse of agrochemicals since application is targeted. These technologies have the potential to reduce the overall environmental impact of agricultural operations, specifically on soil degradation and erosion. Predictive modeling coupled with the common techniques of no-till or minimal tillage will lead to reduced soil disturbance and conserved soil structure and soils with higher organic matter content. These practices reduce erosion, enhance water retention, and promote overall soil health (Bruun-Jensen et al., 2024)

SMART can drive decisions in precision agriculture allowing producers to make decisions based upon real time analysis of data for a particular crop and season. Further, being able to collect and analyze data over time will allow producers to understand trends, optimize selection and planting of cultivars, and determine crop rotations. This continuous monitoring and adaptation help cope with changing environmental conditions and resource availability. Water scarcity is a significant challenge in agriculture. Precision farming incorporates smart irrigation systems that use sensors to measure soil moisture levels. This information enables farmers to apply water precisely where and when it is needed, avoiding over-irrigation and conserving water resources (Bruun-Jensen et al., 2024)

Environmental sustainability is central to the agricultural and food plans of many countries including Brazil, India, Japan and across the European Union. The EU is driving much of the global perspective on environmental sustainability and considers Food essential ingredient for environmental sustainability (Figure 1) (Jesus, 2022).

Figure 1. Map of the 24 potential emerging issues identified per segment of the food chain. From: Reimagining the food system through social innovations — European Environment Agency A Jesus. 2022

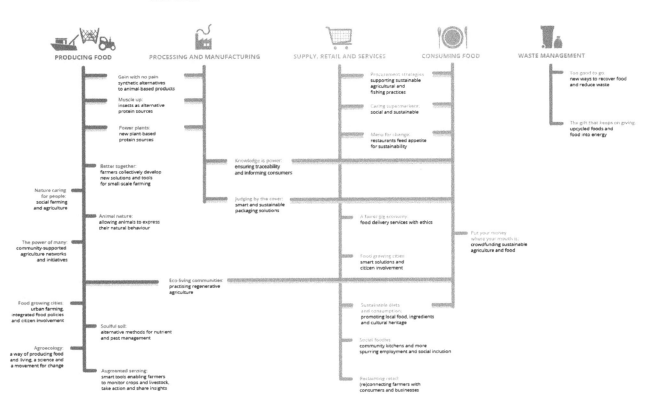

Figure 1. Map of the 24 potential emerging issues identified per segment of the food chain

The tie between food security and environmental sustainability are strong with drought and water availability including timing of snow melt, and rural/urban usage, soil erosion, temperature seasonal shifts and hotter temperatures in key food production areas raising concerns about the sustainability of food production in many regions. High agricultural input due to disruptions in the global market dramatically reduced the availability of fertilizer in 2022 (GAO 2023) and inflationary impacts have kept costs high, fuel costs fluctuate widely and remain high. Further contributing to high costs are monopolistic control of certain inputs such as seeds with four companies controlling over 50% of the world seed supply (Shield 2021). Consolidation in the seed sector has reduced the number of both cereal and horticultural crop varieties available with a loss of 75% of world crop varieties between 1900 and 2000 (Shield 2021) reducing the capability of farmers to respond to shifting factors associated with the growing environment including temperature and rainfall shifts exacerbating drought and shortened cultivation seasons that could be addressed if there was greater seed biodiversity remaining for commercial producers. Genetically altered crop plants that adapt to changing temperature, high soil salinity and drought could alleviate lack of seed biodiversity as can the ability to map growing areas at both a large and very localized level to predict what the growing conditions might be for the upcoming season allowing for a better determination of the best cultivars to raise, when to plant them, and what inputs would be needed for their growth. Promoting a regulatory environment including commodity price supports to accommodate integration of cover crops into rotations, water allocation and usage, and agricultural lending to make this possible will also be required so that agricultural lands can be managed as living ecosystems while providing support for the economic, agronomic and social aspects of such operations. A major shift in the US will be developing a regulatory framework to integrate animal agriculture with crop production systems and greater financial support for regenerative agriculture for commodity and specialty crops. Unfortunately, food safety provisions currently in effect either limit or prohibit these practices. Implementing SMART to improve soil and water monitoring in real time to determine whether conditions are ripe for microbial contamination, real time analysis of agricultural chemical residues in crops and physiological aspects of crops indicative of maturation or stress could improve capabilities to ensure product safety, this is particularly important now with fresh fruits and vegetables.

SMART TRACEABILITY ACROSS THE SUPPLY CHAIN
IoT (Internet of Things)

IoT devices include sensors that can monitor conditions across the supply chain by collecting real-time data on temperature, humidity and moisture content, pH, environmental gas composition during storage and transport of perishable fruits and vegetables has been a major advance for many countries that depend on maintaining optimal conditions for product quality across the supply chain. This is important in tropical and semi-tropical countries across Asia, Africa, Central and South America where cultivation conditions for perishable foods and climate vary greatly and harvest is often under suboptimal conditions, but appropriate technologies could reduce the impact of climate and harvest on food quality. Real-time monitoring can identify potential quality and safety problems quickly, reduce waste and reduce the risk of foodborne illnesses. Temperature monitoring control remains the most important factor for shipments along with controls for plant respiration that make it possible for intercontinental shipment to distant markets by rail and ship. Quality monitoring sensor data integrated with traceability and incorporated into a blockchain provides a means to record characteristics of product quality and the storage environment in an immutable ledger for shipments.

Major advances in traceability, and in the food industry utilization of blockchain is becoming more common and has had the recommendation of regulatory agencies. Blockchain provides transparency across the supply chain verifying important factors besides origin including animal or plant health factors, breed or cultivar and diet or agricultural amendment or chemical usages. Machine learning algorithms are used to optimize supply chain logistics including transportation routes, storage conditions for foods post-harvest and throughout the supply chain as the food passes through multiple hands. Market demand and demand forecasts can be tracked to optimize supply chain (Bruun-Jensen et al., 2024).

Current common features for traceability in agriculture and food processing are barcodes, radio-frequency identification tags (RFID) and interoperable databases in addition to blockchain (Charlebois et al., 2024). Digital traceability provides the ability to rapidly identify foods that may be contaminated facilitating the precise identification of contamination sources, reducing foodborne illness risks. Many countries are integrating digital traceability within their regulatory requirements starting at the source of a food or ingredient and continuing along the supply chain mandating "one step forward one step back" traceability. The US and others have promoted but not mandated blockchain technology. The Food and

Drug Administration (FDA) published the highly anticipated and long-awaited finalized Food Traceability Rule, implementing Section 204 of FSMA (FSMA Section 204) in November 2022. A component of the FDA's "New Era of Smarter Food Safety" blueprint, the rule encourages more investment in digital technology and traceability (Bruun-Jensen et al., 2024). A recent study by the Grocery Manufacturers Association found that high quality traceability can identify the source of a food safety issue or questions of a product's authenticity 90% faster than when a traceability system such as blockchain is not employed.

Blockchain involves a decentralized database system that provides the means for keeping, handling and sharing data from the supply chain from the producer to retailer and potentially the consumer. These systems enhance both product traceability and transparency. Traditional supply chains often utilize a centralized authority or intermediary to manage and validate transactions. Blockchain incorporates a decentralized ledger, eliminating the need for central authority. Each participant in the supply chain has a copy of the entire blockchain, and transactions are recorded across multiple nodes simultaneously creating a transparent and auditable trail. The immutability feature ensures the integrity of the information stored on the blockchain, providing a reliable and tamper-proof record of transactions (Bruun-Jensen et al., 2024). The key characteristics of blockchain—decentralization, immutability, and consensus procedures—make it a trustworthy alternative to traditional centralized databases. Data are kept in blocks in a blockchain system, which are then linked together in a chain, producing a secure and unalterable record of all transactions inside the network (Charlebois et al., 2024) ensuring authenticity and compliance with standards. By creating a "digital twin" of their supply chain, combining traceability and digitalization, companies can play a crucial role in increasing the effectiveness of recalls through the swift identification and removal of contaminated foods from the market (Bruun-Jensen et al., 2024). It is possible to integrate robotics and automation outputs with traceability systems to track and trace products throughout the processing and packaging stages. This enhances transparency and allows for quick identification and resolution of supply chain issues. Automation optimizes resource use, including raw materials and energy, by minimizing waste and ensuring accurate portioning and packaging and adherence to product technical specifications. Adoption of SMART technologies along with blockchain technologies aligns with sustainability goals and reduces the environmental impact of food production (TraceXTech, 2024) which is a major objective of food producers and governmental entities tasked with agriculture and food policy and regulation.

AI (Artificial Intelligence)

Vast amounts of data collected from IoT devices and blockchain records from food industry operations and logistical systems can be analyzed utilizing artificial intelligence algorithms. These algorithms can identify patterns and anomalies providing the means for processors to take a proactive approach to quality control predicting potential issues with contamination or variation from technical specifications or signs of spoilage before these become critical. Among the most important and most popular AI applications for predicting spoilage or food safety is by monitoring and then controlling storage temperature in real time to reduce risk of spoilage and reduce the risk of pathogens from growing. AI makes it possible to decide whether to retain a shipment or redirect shipments for batches to a closer location based upon a risk of quality loss. AI can also play an important role in monitoring news reports and social media for market interest or for signs of potential food safety risks.

Sensors have been included in shipments for decades to track location and to monitor temperature. By supplying continuous data about a shipment, maintaining quality and predicting remaining shelf life at the point of sales can be precisely monitored and controlled. Problem shipments can be immediately identified, making it possible to make better decisions about their disposition.

SMART Applications in Food Processing

The Internet of Things (IoT) employing sensors that are compatible with SMART monitoring are widely used to optimize crop production and improve resource utilization in farming. In food processing, IoT devices are coupled with robotics to reduce the risk to people from repetitive and dangerous tasks, tighten control parameters thereby increasing efficiency and reducing waste. Machine learning models enhance the capability of SMART devices to determine when maintenance is needed and to monitor product and process attributes in real time allowing for adjustments that ensure food safety and quality. AI integrated computer vision systems equipped are used to for quality control and grade inspections of fruits, vegetables, nuts, seafood and meats assessing size, color, shape or configuration, presence of contaminants such as shell and surface and internal defects. These vision systems can be integrated with processes such as blanching or dehydration to ensure heat transfer is adequate to inactivate enzymes to maintain color, eliminate microbes and hit target moisture levels so that these foods are safe and meet consumer expectations for flavor, texture and appearance reducing the risk of either under or overprocessing.

Incorporating smart sensors to in-line processing systems to predict risk of microbial presence such as the presence of organic matter or ATP (adenosine triphosphate), presence of proteins (allergens), presence of residual cleaning chemicals, or wear and tear on metal, plastic and rubber parts that could be a source of physical contaminants, makes real time control of contamination risk possible. Most of our process control instrumentation for temperature, flow rate, fill, volume or weight, pressure, pH and chemical concentration (conductivity, ions) are now fully digital IoT devices that can be linked to integrated control systems to provide tighter process control and adherence to technical specifications and with this greater efficiency and reduced cost. Improved sensors with reliable signal accuracy at reasonable cost have led to improved process control specifically for pH and temperature sensors that can drift. These monitoring technologies improve efficiencies in energy and water use addressing concerns consumers may have about environmental sustainability. Instrument controlled clean in place systems are replacing timed cleaning processes and can have greater effectiveness and dramatically reduce water, chemical and energy use. Sensors with AI capabilities can provide diagnostic information on the condition of the sensor as well as detect more quickly system anomalies that could affect process effectiveness and pose a risk to the safety or quality of the food being produced. These features improve food safety since access to digital sensor data can be shared. Sensor output can be integrated with traceability programs to improve tracking of ingredients sourcing and use and product production within a facility and tied to customer systems as needed to show traceability.

An increased use of robots or co-bots that assist humans with repetitive tasks that are prone to error, such as sorting, cutting, peeling, packing and labeling with high precision and speed having the potential to reduce food safety risks. The integration of robotics and automation in food processing and packaging is transforming the industry, improving efficiency, adherence to technical specifications and scalability (TraceXTech, 2024). Food processors have utilized visioning systems for over 20 years to detect defects and ensure quality and consistency and improving the integration of machine vision into process automation provides a means to reduce waste. Robotic systems employed in the packaging and handling unit operations of filling, sealing, labeling, casing and palletizing are becoming more common. Robotic arms can rapidly and precisely pick and place items into packaging at high speed and with the uniformity and appearance required to meet customer specifications (TraceXTech, 2024).

SMART and Novel Food Development and Personalized Nutrition

Smart food technology will also be driven by the adoption of demands created by novel foods and the novel processing methods required to make them. The market demand for these new foods responds to environmental or ethical, primarily animal free, concerns. Foods are now engineered from the molecularly identical ingredient up. Such products include "cow free" milk" a primary constituent being β-lactoglobulin made by genetically engineered yeast fermentation. Claims are that milk made by fermentation processes uses less water and energy and creates fewer greenhouse gas emissions that milk from cows. Similar claims are made for cell-based meat in addition to the potential to reduce food borne illness from transmission of pathogens, most importantly antibiotic-resistant pathogens from meat to people. Cell-based or lab-grown meat and seafood is a rapidly growing area that employs sophisticated biochemical and material science processes to create muscle cells in bioreactors that are then assembled into structures that resemble whole muscle tissue that includes both muscle cells and fat cells or adipocytes. Employing scaffolds and 3-D printing, common in other industries but relatively new to food processing where the scale and demands of cost-effective scale-up makes this form of food production different in scope than pharmaceutical production and tissue engineering for regenerative medicine. Both FDA and USDA realize that more technically advanced and SMART analytical methods for safety and traceability are required to meet demands of the emerging markets for personalized nutrition and medical foods.

SMART AND GOVERNMENT STRATEGIES AND PROGRAMS
The OECD Countries and European Union, Generally

Agriculture and food processing sectors understand that adopting SMART automation and other new technologies are essential to remain competitive and promote innovation. Many of the OECD countries are supporting research into the use of digital systems including those important to food safety specifically traceability. In a recent study by (Charlebois et al., 2024) an investigation of programs in the 38 member countries of the OECD: Austria, Australia, Belgium, Canada, Chile, Colombia, Costa Rica, Czech Republic (Czechia), Denmark, Estonia, France, Finland, Germany, Greece, Hungary, Iceland, Ireland, Italy, Israel, Japan, Republic of Korea, Latvia, Lithuania, Luxembourg, Mexico, Netherlands, New Zealand, Norway, Poland, Portugal, Slovak Republic, Slovenia, Spain, Sweden, Switzerland, Turkey, the United Kingdom, and the United States. Their analysis extended to the European Union, focusing on the 23 member states who are OECD members.

OECD countries are focused on advanced technologies and digitalization to improve food safety, traceability and regulatory conformity (Charlebois et al., 2024). EU countries are evaluating a multifaceted food system encompassing various initiatives, platforms, and technologies that are aimed at ensuring the safety of all imported and exported foods and to reduce the risk of transmission of animal diseases. The primary system used across the EU to improve traceability of live animals and animal foods is TRACES (Trade Control and Expert System) mandatory electronic certification. This system ensures that the health and safety regulations of the member states are met with the food sold and exported and provides an efficient and transparent logistical control system showing the integrity of the overall food safety system for animal foods. The information from TRACES provides a vast database of information that is critical for both the public and private sector by offering up-to-date information on food safety and in addition providing data for the EU Food Security Index. Through the Joint Research Center of the European Commission a big data platform that incorporates traceability data is being developed to provide a centralized resource for the agriculture and foods sectors (Charlebois et al., 2024).

To advance the EU's sustainable food systems it has launched a Farm to Fork Strategy that outlines a comprehensive roadmap that will improve food traceability and transparency across the food value chain. One aspect of this is the AgriData Space collaboration to build a decentralized digital platform that will promote data sharing. The General Food Law (Regulation (EC) No.178/2002) provides the legal framework that governs digital traceability in the EU and specific principles and requirements food and feed industries are to follow regarding food traceability. This law along with Regulation (EC)No. 852/2004 requires that food producers and sellers be able to identify and trace each participant and lot of products and product attributes from the farmer through to retail and maintain digital records that make prompt and precise information retrieval possible in event of a food safety incident. (Charlebois et al., 2024). Clearly, SMART and blockchain will be critical for making this system efficient, effective and transparent.

Adoption of SMART Food Safety in Several Countries

SMART is a focus of food safety initiatives in many countries with a summary of some of the major initiatives provided here (Table 1). Each of these initiatives incorporates digital traceability as a primary component, and along with these strategies to use SMART to improve food safety management programs, specifically for chemical and microbial testing and data sharing; with recommendation for process monitoring and how SMART can provide improved preventive controls. SMART is also a factor in food defense, food adulteration control and with programs to support environmental and agricultural sustainability.

Table 1. International Initiatives in SMART Food Safety

USA	Smart Food Safety	New Era of Smarter Food Safety Blueprint | FDA
EU	Food 2030	Food safety in the future EU budget (2021-2027) - European Commission (europa.eu)
Japan	Smart Food Value Chain	Japan's Smart Food Value Chain: From Consumer to Agri-Food Industries | FFTC Agricultural Policy Platform (FFTC-AP
China	National Food Safety Standard Legislation (2023)	China's Food Security: Key Challenges and Emerging Policy Responses (csis.org) [Updated] China Unveils National Food Safety Standard Legislation Plan for 2023 | ChemLinked China: New and Amended National Food Safety Standards | USDA Foreign Agricultural Service
Canada	Next Generation of Food Safety	Renewing the Food Policy for Canada: Towards 2030 – Food Secure Canada Health Canada 2022 to 2023 Departmental Sustainable Development Strategy Report - Canada.ca Next generation of food safety - Canadian Food Business
Singapore	Smart Nation	Food Regulations - Singapore Statutes Online (agc.gov.sg) Smart food: novel foods, food security, and the Smart Nation in Singapore (researchgate.net) Science of safety: Singapore's Future Ready Food Safety Hub outlines novel foods, agriculture and aquaculture focus (foodnavigator-asia.com) Food Safety and Security Bill in the works, will give more clarity on novel foods | The Straits Times
Brazil	Vison for Global Food Security	Brazil 2050: A vision for global food security - Atlantic Council
New Zealand	Food Safety Strategy	New Zealand Food Safety Strategy | NZ Government (mpi.govt.nz) Food Standards Code | Food Standards Australia New Zealand
Australia		Standard 3.2.2 Food Safety Practices and General Requirements. pdf (foodstandards.gov.au)Safe Food Australia 2023 - updated guide to Food Safety Standards | Food Standards Australia New Zealand

India		Establishing a regulatory framework for a sustainable food future in India: DBT's leadership and foresight - GFI India (gfi-india.org)
		Future-forward food safety: Leading the way in India's next gen (foodinfotech.com)
South Korea		Ministry of Food and Drug Safety>Information>News&Notice>News&Notice> View Details	
		Ministry of Food and Drug Safety (mfds.go.kr)

Australia

Australia is a major exporter of beef, lamb, grain and fruits and vegetables at >80B USD per year, and a leader in traceability and the use of data science and digitalization to enhance supply chain efficiency. Australia's Department of Agriculture, Water, and the Environment have several initiatives to enhance digital traceability in the agricultural sector improving traceability and food safety in its regulatory programs (RegTech) and Agrifood Connect Trace2Place that permits traceability of meat across the supply chain in real time (Charlebois et al., 2024). The country has implemented a system called eCert that allows government agencies to internally exchange certificates on imports and exports. Meat & Livestock Australia Limited, which has adopted an Australian AgriFood Data Exchange to enhance compliance in the supply chain through a cloud-based platform. Similarly, from these and other data Australia has developed a risk assessment tool for red meat, dairy and horticultural supply chains. Australia and codified food safety management tools (Standard 3.2.2) a code shared in part with New Zealand that addresses aspects of food hygiene and safety in addition to gene edited foods along with traceability.

Brazil

Brazil has a thriving agricultural sector and is a major food exporter with 113B USD exports of soy, beef, sugar, grain and coffee and 288M USD in processed foods. The country is focused on food security, food safety, particularly of novel foods, and environmental sustainability of its agricultural sector and food waste.

Brazil has a vision for global food security and the role it can play in solutions to meet the food needs of a growing population, and how changing demographics and demands for protein heavy diets will affect the country's agricultural development. From its recent history, Brazil understands how environmental degradation, economic instability and shifting geopolitics can impact agricultural productivity and global food supply chains. Brazil promotes novel foods and ingredients, their safety, and regulates their use (RDC 839/2023 Brazilian Health Regulatory Agency (Anvisa). Brazil will play a major role in addressing global food security and addressing the lack of food faced by 11% of what world's population, in recent years food security has improved significantly in Latin America due in part to Brazil's production of grain, oilseeds, sugar, meat and animal feeds and because of the country's ability to increase production of grains and oilseeds relative to other major producers including China, India and Russia, Brazil will play a major role in meeting global food demands in the next decades. Despite this advantage Brazil has much degraded land.

Policies in Brazil on innovation, infrastructure and investment for productivity gains will be important to global food security and the country is focused on how technologies for food production and distribution and transformed itself into an international leader starting in the 1960s to become a leading food exporter by adopting efficient and sustainable production practices, but issues remain. Brazil imports much of the technology it uses and is implementing no-till practices and precision agriculture and on-farm robotics on large corporate farms. A skilled workforce, specifically individuals capable of responsible resource management and those who can adopt technological innovations is a crying need. Rural development is held back because of poor transportation infrastructure, lack of sufficient modern food storage capability and unstable energy systems. Brazil is a large country with an extensive river system but uses only 30% of its navigable waters for transportation of people or goods. Brazil's road and rail transportation systems are inadequate. Access to capital for small and medium sized agriculture producers and smaller food processors so that they can invest in modern equipment and adopt technologies that would lead to increased productivity and profitability, supporting land restoration and sustainable practices.

The Brazilian government provides agricultural subsidies, targeted private sector investments and microfinancing to expand agricultural operations incentivizing grain and legume production. The country is using funds to promote crop diversification and adopt restorative agriculture, and other sustainability practices such agroforestry and reversing deforestation, carbon capture, integrated pest management, conservation tillage, intercropping and integrated crop-livestock systems in addition to water resource management. This support includes funding of technological innovations in sensing, monitoring and data science that could increase grain and soybean production each by over 30% with no reduction in forests and at lower carbon emissions. Brazil has a competitive advantage over North American and European temperate climate producers by being able to produce two and sometimes 3 crops per year, however to be successful, SMART will need to be built into these initiatives if they are to have the positive impact on global food security that the country seeks without further degradation of their critical forest and water ecosystems by adopting remote sensing, robotics, automation and artificial intelligence applications. Brazilian agriculture has sufficient water resources to expand irrigated crop area ten times but has overused ground and surface water resources in some regions; areas of the country have also been affected by drought and increased aridity. Because of this opportunity, Brazil should prioritize adoption of the best available technology to develop irrigated systems at both the local and national levels.

Canada

Canada is focused on smart food safety. (Charlebois et al., 2024) notes that Canada is heavily invested in a robust and effective regulatory system and is focused on expanding digitization in food safety risk management and inspection with the objectives of creating benefits for producers, importers and exporters including digital traceability but efforts to build out a federal system are currently underfunded. The country is committed to adopting advanced technology, big data, digital tools, and automation that can facilitate rapid, accurate, sensitive monitoring of the food supply chain in a user-friendly manner pairing food safety practices such as HACCP with IoT and SMART, traceability, whole genome sequencing and rapid analytical methods and sensing technologies to improve food safety without limiting food choices to manage safety remembering that risk analysis and risk mitigation must take into consideration cultural values and consumer preferences. **Renewing the Food Policy for Canada: Towards 2030 – Food Secure Canada**..

Canada participates the GS1global system for food safety but has yet to promulgate regulations in this area, although traceability requirements in the US will likely drive regulatory adoption. Canada exports roughly a third of the food and beverages it produces and is 5th largest food exporter in the world (https://foodpolicyforcanada.info.yorku.ca/backgrounder/problems/reliance-on-exports) with $28.3 B USD in 2023.Canada is adopting a food safety blueprint similar to that of the US FDA emphasizing 4 core elements of tech-enabled traceability, smarter tools and approaches for prevention and outbreak response, new business models and retail modernization, and food safety culture and redesigning regulations to reduce impediment preventing or delaying from bringing innovative products to market while protecting health and safety. Canada is also committed to new paths to market entry for supplemented and novel foods by providing greater flexibility to respond to market forces and advances in science and technology. **Renewing the Food Policy for Canada: Towards 2030 – Food Secure Canada**

Canada is positioning itself to be a leader in SMART food safety technologies emphasizing that more data on foods can be collected with advancement and integration of Internet-of-Things (IoT) devices and these data collected from monitoring devices then analyzed in real time recognizing that the availability of fifth generation (5G) networks that provides significantly more bandwidth than older networks provides for real time data transfer with low latency with fewer local restrictions. SMART sensors that detect temperature, humidity, gas composition and levels in a food container headspace or food storage environment provide continuous monitoring of factors impacting food safety and quality such as freshness and when deviations are noted, prompt corrective action can be taken. Cloud computing will play an important role in the capability of the food industry to manage complex computational models and employ machine learning algorithms for decision making without the need to maintain costly hardware. Whole genome sequencing coupled with bioinformatics analysis has revolutionized food safety testing making it possible to determine the identification of a microbe in a matter of minutes or hours rather than days. Rapid analyses for food components and chemical or microbial contaminants has advanced quickly but the major limitation of employing these technique in food analysis remains the ability to separate and recover the analyte from complex food matrices and to accurately detect contaminants without interference from other compounds. Extraction methods for analytes in low concentration in solid foods such as meats and cereal products remain problematic. The development of antibody, aptamer and molecularly imprinted polymers that have binding sites for target molecules make them highly selective reducing matrix effects when coupled with solid phase extraction and selective membrane filters. Renewing the Food Policy for Canada: Towards 2030 – Food Secure CanadaHealth, Canada 2022 to 2023 Departmental Sustainable Development Strategy Report - Canada.ca, Next generation of food safety - Canadian Food Business

China

China is concerned about both national food security and maintaining its position in the global food export market and rebuilding the food economy following COVID. China is a major exporter of seafood, fruits and vegetables, tea, processed foods and beverages and cereal grains with a value of $76.4 B USD in 2021 and with ensuring the safety of imported foods. China imports $98B USD in 2022 with much of this being meat and poultry and animal feed components to meet the demand of the expanding urban middle class for more animal protein and dairy foods and higher quality food generally. China has faced numerous scandals with contamination of meat and dairy with the devasting baby formula (dried milk) incidents starting in late 1990s, widespread adulteration of oils, bakery products, dairy foods, soy sauce and various beverages. In the animal foods realm, cases of species substitution, and misuse of animal drugs in domestically produced poultry, pork and fish. China's middle class with be 50% of its population by 2025 and meat consumption has tripled since 1990. Proposed regulations to ensure national food security and food reserves, protect farmland, and maintain an effective logistical system for food distribution and storage are being implemented. To achieve these objectives China seeks to modernize its domestic agriculture sector and to incorporate strategic AI technology starting with monitoring agricultural inputs and for pest control a significant issue for food quality and safety in addition to microbial and chemical contamination. The country is investing in research into GMO and gene editing of food crops and in alternative or novel foods. China is one of the few countries focused on a blue granary to establish mariculture, exploiting international deep-sea fisheries and developing both coastal and deep-water aquaculture both of which have unique challenges that SMART could address both for production and safety, to optimize feeding, water quality and to manage fish populations and fish health.

China is facing a difficulty with increased agricultural productivity along with a labor shortage in agriculture, transportation and logistics and automation. SMART technologies could address to some extent and is making investments to improve agricultural productivity by reclaiming farmland, improving irrigation, and investing in water saving technology and water diversion projects. In addition, China is purchasing or leasing 6.5M hectares in Asia and Africa for agriculture, forestry and mining activities. China is building a land grain corridor to Russia for grain imports. Fragmentation of agricultural land has hindered technological advancement and market standardization including compliance with international food safety requirements, and sector coordination, in addition to incentivizing small holders to raise cash crops at the expense of grains. Unfortunately, this move to further coordination and centralization of agricultural production will reduce product diversification but may make it easier to adopt modern production technologies due to economies of scale. This presumes, however, that there will be at least to some extent an open market.

The regulatory climate for food safety in China lacks transparency. The incentive structure within the Chinese bureaucracy that encourages misrepresentation of data by promoting people who report good news instead of facts and undermines the development of a robust food safety culture. Corruption at all levels of government creates a lack of trust in the food safety system in domestic and international market. The first comprehensive food safety law in China was adopted in 2009 and requires food processor licensing and a safety review of food producers and of the foods entering the market. This law established national food safety standards normative of prevailing international ones for storage and distribution, ingredients, packaging, sanitation, labeling and product testing. A new food law in 2022 addresses increasing demand from consumers for high quality safe food including GMO and alternative proteins and with this a focus on technology to improve efficiency in the agriculture sector and to provide digital platforms for sharing information, traceability and supply chain management.

Columbia

Columbia shows that with the right incentives, digital technology can be adopted. Despite a limited network infrastructure across its rural areas with only 30% of households having internet, Columbia has a highly effective traceability for its high value cash crop, coffee across the entire supply chain using digital platforms. Columbia produces about 1/3 of the worlds coffee and government-industry partnerships such as the Coffee Information System (Sistema de Información Cafetera, or SICA), run by the non-profit organization Colombian Coffee Growers Federation, provides information-based database to profile farmers and their products offering transparency in the coffee production supply chain to importers and other key stakeholders with transnational documentation to support each transaction across the supply chain. Blockchain based traceability has not been fully implemented because of the difficulty of farmers sharing information due to lack of a uniform system or platform for collecting data and cultural challenges impeding the information integrity (Diaz, Rojas and Moncayo (2021). This example shows that adoption of SMART for high value products raised by small shareholders, in this case coffee, but also tea, cacao, spices, botanicals and certain seafood are possible when there is coordination through a growers' cooperative to make such an effort possible.

India

India is a leading agriculture producer and is aggressively adopting new technologies for production and food safety as a strategy for increasing international markets for their food products and improving food security. Like Columbia, India is encouraging producers to integrate emerging technologies into their supply chains and blockchain is being implemented to trace tea and spices to international markets to ensure transparency and safety building consumer confidence about the safety and authenticity of these high value foods reducing the risk of adulteration and economic fraud. India recognizes that food adulteration and fraud are significant challenges in the agricultural sector and are encouraging adopting of SMART technologies to address this problem. Farmers gain access to a trustworthy system that fairly records their use of agricultural inputs, crop characteristics, volume and price of their products helping them to obtain better prices and market recognition for quality. QR codes as part of food packaging and mobile apps are becoming more popular as a means of raising consumer awareness about food safety, instructions for product use and comprehensive information about the food source, composition and handling.

India is taking international leadership on alternative proteins and with cultivated meat and novel 'smart' proteins including development of a rational regulatory framework and progressive rulemaking and risk management at the federal level addressing existing regulatory bottlenecks preventing a path to market for novel proteins and safety monitoring. These novel foods include cell culture-based meat, fish, and dairy, plant-based protein alternatives, fermentation derived proteins, 3-D printed foods, seaweed, microalgae, and insects. Fermentation products and cell-based meat require premarket approval whereas plant-based meat analogs, egg and dairy foods that use standardized ingredients in an innovative way that does not require extraordinary safety evaluation (a proprietary food) are not considered to be novel foods requiring premarket approval. However, with proprietary foods it is advisable to require guidance through a preapplication consultation process with the Food Safety and Standards Authority of India (FSSAI) like processes in place in Singapore and the USA. Currently the number of novel protein food companies in India is small at about 15 but significant growth in this sector is anticipated with growth as an ingredient supplier for international companies as well as domestic producer of novel foods. FSSAI has approved 3 novel protein ingredients: Perfect Day's non-animal whey protein derived from precision fermentation, ACME's mycoprotein recovered from *Fusarium venenatum* and Reliance's phototropic algal biomass derived protein powder. A regulatory framework that is less stringent and lower cost manufacturing models than that of the pharmaceutical industry but that still maintains high product safety will be required for novel proteins will result in a unique set of compliance requirements that are more stringent than conventional food production but less stringent than pharmaceutical production defining this as a priority under the India Department of Biotechnology (DBT) through the Fostering High Performance Biomanufacturing initiative. India also recognizes the need for advanced technologies specifically biotechnology and SMART to combat climate change, improve food security and meet the nutritional needs of a growing global population.

Japan

Japan has stringent market requirements and quality standards for food safety tracing the use of digital technologies to the bovine spongiform encephalopathy, or "mad cow disease" outbreak in the late 1990s and from this developed a unique 10-digit ear tag as part of an Individual Identification Register system for beef cattle that tracks birth date, breed, health care and shipment details when exported. (Charlebois et al., 2024). This traceability system could be expanded to other foods and the government since 2001, in collaboration with private companies, is making efforts to integrate advanced information and technology in the agriculture sector through programs in the Ministry of Agriculture, Forestry, and Fisheries (MAFF) (Charlebois et al., 2024) although insufficient resources have been available to fully implement digital traceability in conjunction with RFIDs and barcodes. Consumers can trace many foods through the supply chain at point of sale through QR codes and apps accessing information on origin, safety and use of genetically modified ingredients. Japan is committed to integrating physical space and cyberspace to create a smart food value chain by integrating information and communication technology (ICT) with AI. Japan's Smart Food Value Chain: From Consumer to Agri-Food Industries | FFTC Agricultural Policy Platform (FFTC-AP) and sees this as a means to promote both food safety but also environmental conservation goals. Standardization of terminology and platforms will be an important component of these efforts to optimize the value chain by improving productivity and quality, reduce waste and cost and meet market needs, match needs. Information sharing and logistics across value chain. Standardization of terminology in quality evaluation. Use AI and ICT to optimize value chain. Japan is creating an AI based agricultural platform promoting an integrated SMART breeding, production, processing and distribution system for domestic and export sales. This strategy is needed to address labor shortages in the agricultural sector including robotics including self-driving "agribots, " and smart greenhouses that reduce the risk of food contamination. Japanese producers are extending the use of SMART to processing and distribution using this to predict quality and product safety through the use of sophisticated spectroscopic sensor to predict ripeness and other quality factors for example, reduce waste from deterioration and minimize energy usage. The WAGRI agricultural data collaboration platform was designed to integrate sensor based real time data on climate, soil, water and input usage, location into AI based models that assist with the coordination of plant growth with market demand and with this to maximize quality and product safety.

Republic of Korea

Korea is adapting blockchain technology to assure food safety starting with traceability of beef to track safety and credence attributes (Kim 2009) and has the capability to trace beef through the entire value chain and is adapting 5G blockchain and SMART to other foods through private partnerships between food companies and telecommunication companies (Charlebois et al., 2024). Scannable QR codes are available for consumers at point of purchase for many foods. Korea is one of the first to develop a global digital traceability system specifically for halal foods. Korea is committed to technological advances for import inspection implementing SMART HACCP that automatically identifies product labeling and safety information at point of inspection with mobile phone that is tied into quality inspection programs and data for manufacturers. This joint program between the Ministry of Food and Drug Safety and the Korea Chamber of Commerce that can very quickly determine if a hazardous food is on the market and prompt a recall using and internet-based system (Figure 2). Korea is committed to strengthening regional collaborations across the Asia Pacific with Australia, Chile, China, Indonesia, Malaysia, New Zealand, Philippines, Singapore, Thailand and Vietnam to integrate and harmonize food safety programs and shift to a digital food safety management program. This will provide greater market access and expedite food trade.

Korea is also committed to amending its regulatory framework to accommodate cultivated meat and to provide regulatory free zones to foster cultivated meat R&D.

Figure 2. Traceability in Korean Food Sector

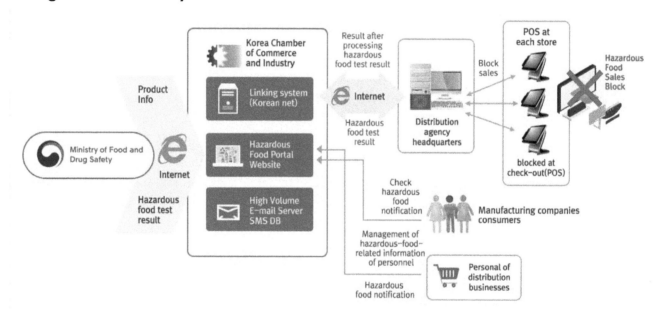

Cite as: Ministry of Food and Drug Safety>Our Works>Food>Food Safety | Ministry of Food and Drug Safety (mfds.go.kr)

New Zealand

New Zealand is moving towards a fully digital system for food traceability from production through to consumer sale driven by food safety concerns. The beef supply is fully traceable using barcode, QR code or RFID technologies with mandatory tagging in 2012 through the National Animal Identification and Tracing Act with the goal of completely shifting to a blockchain based system. New Zealand is promoting a strong industry-based food safety culture to be able to ensure that foods for domestic consumption and export are safe and meet technical specifications. Agriculture is about 15% of New Zealand GDP and food constitutes about41 B USD and are 66% of total merchandise exports. The country is committed to the integrity of its food safety system and is supporting regulatory standards to maintain and increase value of exports through regulatory compliance to ensure New Zealand food is trusted and recognized by overseas and domestic markets as safe and authentic making it easier for businesses to export upgrading certification systems that provides flexibility to New Zealand producers while still ensuring safety and adherence to technical specifications. Annual programs for recall practices and development of guidance to ensure standards of export markets are met are key provision of food safety regulation. New Zealand is prioritizing SMART and technologies that can assist with risk assessment and risk management. In addition to safety, consumers' concern with nutrition, food security, animal health and welfare, climate change and sustainability are driving regulatory programs within the country.

Singapore

Singapore is a highly trusted producer of processed foods and embraces digital technology in the food space as aligning with its vision of building a "smart nation." Singapore is a leader in novel foods and with the standardization, regulation, intellectual property, data-gathering, and training needed for these foods as well as will more traditional products.

Singapore is also focused on food security and through the Food Safety and Security Bill will make it possible to produce more staple foods. Emphasis on food safety and food fraud have made Singapore a leader in adopting rapid analytical methods and AI for screening conventional and novel foods for contaminants critical for food security in a nation that imports 90% of its food. Singapore as a goal of producing 30% of its food by 2023 by adopting technologies for production of cell-based meats, aquaculture production of marine animals and plants, microbial fermentation for food macro and micro nutrients, applications for insects and insect protein and incentives for urban farming and this will only be possible by utilizing SMART to maximize both production efficiency and product quality and safety. Singapore has had a liberal policy for granting regulatory approval for novel foods and is the first country to approve cell-based chicken.

United Kingdom

The United Kingdom has several initiatives in digital traceability (Charlebois et al., 2024) covering the supply chain to the retail and consumer levels for beef, pork, wine imports and foods with multiple ingredients such as sandwiches utilizing smart tags for localization and temperature monitoring to predict shelf life and provide consumers and retailers information about product source and credence attributes. Other initiatives use blockchains and integrated DNA based technologies to address food safety challenges. Recent regulations for imports and exports are based on digital platforms across stakeholders in the supply chain, such as the Defra Digital Assistance Scheme that allow electronic sharing of export health certificates for livestock (Charlebois et al., 2024). The United Kingdom is committed to food safety with a current focus on the safety of alternative proteins and foods produced by fermentation and tissue culture. FSA has issued guidance for businesses using cell-cultivated products.

United States

FDA is committed to improving the safety and quality of the US food supply and as part of their proposed Smarter Food Safety Blueprint have initiatives in technology enabled traceability, data management for food safety and smarter tools for food safety outbreaks, smart technologies for novel foods, technological advances to improve nutrition, strategies for developing a more sustainable food supply, new business models and retail modernization and how SMART food technology can improve the food safety culture. New Era of Smarter Food Safety Blueprint | FDA.

The United States is committed to digitizing the food supply chain (Charlebois et al., 2024) launching a 10 year plan in 2020 for a Era of Smarter Food Safety that has the goal of implementing electronic traceability through domestic and international supply chains for high risk foods that are on the National Food Traceability list under the traceability rule recently implemented as a regulation final rule on food traceability under Section 204(c) the FDA Food Safety Modernization Act (FSMA) New Era of Smarter Food Safety Blueprint | FDA. This will drive electronic traceability for sectors of the industry that have not yet moved from paper-based systems. The FDA requires that foods subject to recall be identified within 24 hours and with SMART the goal of tracing food within a matter of seconds will be feasible. **Supply Chain Transparency: Making the entire supply chain transparent to identify and mitigate risks.** Harmonizing tracing activities to support interoperability is an FDA priority along with finding electronic solutions achievable for companies of all sizes to achieve end to end traceability across the food system. New Era of Smarter Food Safety Blueprint | FDA.

FDA technology enabled traceability involves a **Product Tracing System** for internal product tracing (PTS) that will receive and analyze traceability data from industry to make recalls more efficient and enhance existing foodborne outbreak response processes. Developing **Low- or No-Cost Food Traceability** will encourage small and medium sized businesses, technology companies develop traceability hardware, software, or data analytics platforms that are low-cost or no-cost to the end user.

The Agency's initiative for management systems and smarter tools for foodborne outbreak would improve predictive analytics for outbreak response through the development of AI based technologies including artificial neural netrworks, machine learning and generative pretrain transformers (GPTs) for root cause analyses and predictive analytics. With this would be training for inspections, outbreak response, and modernization of recall protocols for federal and state partners **Domestic Mutual Reliance**. One initiative that is currently in place is the **Seafood Artificial Intelligence Learning Pilot** that uses AI and machine learning (ML) to strengthen import screening and to ensure that seafood meets US regulatory and safety requirements. **Remote Regulatory Assessments** is another initiative to improve the safety of the food supply at foreign and domestic food facilities that saves time and resources. Along with this is a voluntary pilot program **Third-Party Food Safety Audits** and integration of federal, state, international and university laboratories into their **GenomeTrakr Network** allowing for the submission of genetic sequences from parasites, pathogens and viruses isolated from foods.

Artificial intelligence has the potential to revolutionize food systems by improving food safety, automating and accelerating production, quality control and inspection tasks and to develop new products that improve nutrition, support healthy aging and are sustainable (https://www.foodnavigator.com/Article/2024/09/16/AI-Unlocking-hidden-opportunities-for-food- and-beverage) FDA recognizes that companies with access to large amounts of historical data may be the first to benefit from development of AI models to develop novel foods and ingredients. Advances in AI models from academic research in microbial genetics and the digestive microbiome will lead to new probiotics that could improve gut health along with development of new foods and ingredients from plant-based sources or those from fermentation that can substitute for dairy or meat based foods, and proteins and lipids with novel functionality or sensory characteristics. Optimizing processes and improving cohort selection for clinical tries to identify new ingredients for both market and regulatory approval is an area where AI has great potential through models that screen alternatives and reduce the time it would take for human expert review prior to market and consumer testing similar to the tack FDA is taking with the approval of clinical applications for radiology, cardiological and neurology products.

FDA recognizes that an important aspect of SMART in evaluating new business models and exploring ways to modernize its response to the revolution in the retail sector to improve food safety in restaurant, stores and e-commerce and omni-channel food distribution. Food sales have moved beyond conventional stores to sales online and via mobile devices for consumers and business-to-business interactions. Agency food safety programs have not been able to keep up with the innovation in food safety and the emerging issues that the new distribution channels for perishable foods present and what the best means is for regulation of food safety in these new spaces of direct-to-consumer sales, grocery delivery and concierge services. Food companies are becoming nimbler to customer demands for product availability, pricing, health and wellness benefits, sustainability and supply chain transparency and are migrating to the cloud to be more responsive to consumer demands and personalize customer engagement. Building a new and modern food safety culture with the USDA and state partners but also with businesses that formerly were not considered to be part of the food production system will require insights from businesses to ensure food safety.

European Union

The European Union remains focused on food safety but has making environmental sustainability and societal objectives a requirement for food business operating in the EU. Corporate Sustainability Reporting Directive (CSRD) with companies having to annually report how they will meet EU Sustainable Development Goals (SDGs). 2024 Food Industry Trends: Tech and Sustainability Insights including methods to combat deforestation and promote sustainable sourcing practices. The EU Deforestation Regulation.

Smart food safety technologies have been roped into the European Green Deal (Jesus, A. 2022). Reimagining the food system through social innovations — European Environment Agency requiring that greenhouse gas emission loss of biodiversity and social impacts on human health from malnutrition, diet related obesity and chronic disease be addressed along with safety and quality. This will involve a fundamental shift in lifestyle and patterns of consumption and production food methods with young Europeans demanding changes in the food system to make it more sustainable (EIT, 2021). Menu for change: Gen Z demand overhaul of food system, European Institute for Innovation & Technology, accessed 18 July 2024. Animal welfare and a drive towards sustainability is placing an emphasis across the EU on plant based proteins are likely to be a center piece of this strategy but this will require an evaluation of the safety of new processing technologies such as 3D food printing, CRISPR-Cas9 (Clustered Regularly Interspaced Short Palindromic Repeats) gene-editing of crops in response to environmental stress, to improve nutritional content or reduce spoilage risk. Food coatings derived from plants or by fermentation have the potential to reduce food waste as do new more sustainable food packaging systems, but safety of these need to be evaluated in terms of both the components in the packaging as with regard to how food spoilage and pathogen growth conditions in the food during storage, distribution and consumer handing may change Food 2023. This research and innovation policy framework supports the transition towards sustainable, healthy and inclusive food systems. It aligns with the European Green Deal, Farm to fork and

Bioeconomy strategy. Associated with this goal of a sustainable food supply is an emphasis on food fraud and supply chain transparency. Food safety, animal welfare, fight against food waste or sustainable use of pesticides are part of the new Common Agricultural Policy objectives.

Alternative farming methods such as vertical farming and hydroponics, urban farming and robotic or 'smart' kitchens provide new opportunities to increase food production nearer to the point of consumption but could also pose food safety challenges. SMART is a key component of these relatively new food production methods and being able to take more holistic approach to food safety by the regulatory agencies involved will be paramount. Food 2023. This research and innovation policy framework supports the transition towards sustainable, healthy, and inclusive food systems. It alighns with the European Green Deal, Farm to Fork strategy and Bioeconomy strateguy. The EU #Safe2EatEU campaign has 17 participating countries for 2024: Romania, Czechia, Hungary, Greece, Estonia, Croatia, Italy, Latvia, Cyprus, Slovenia, Spain, Luxembourg, Slovakia, Austria, Poland, Portugal, and North Macedonia stressing the importance of a balanced diet, the safety of supplements, additives and novel foods, and the validity of health claim labeling Safe2Eat campaign. Food fraud remains a concern across Europe as are the risks from globalization of the food supply and the inherent food safety risks. Food Fraud: When a Food Product Is Not What It Seems (fooddocs.com). Food fraud includes mislabeling and tampering with a food to make it appear to be better or different from what it is, substituting a lower value ingredient for a higher quality one, diluting a food – common with juices or alcoholic beverages, producing a counterfeit product, adulteration by adding inferior or harmful ingredients, counterfeiting and dilution of food products. A counterfeit is a fake version of branded product sometimes at a lower price. Dilution reduces quality by adding water or fillers and is common with juice and alcohol. Unique to the European market, selling a product beyond its use by date. **Food Safety in the Future EU Budget (2021-2027): This initiative focuses on preventing food waste, combating food fraud, and supporting sustainable food production and consumption**. The globalization of the food supply can make adulteration and food fraud more difficult to detect as supply chains become more complex and with this the greater opportunities to commit fraud. Preventing fraud requires proactive interventions, rigorous and sophisticated monitoring and transparency in trade practices, all of which would benefit from implementation of SMART features in food production and trade. Geopolitical conflicts are increasing the risk of food being used as a tactic to cause harm and reduce consumer confidence Proactive Food Defense Plan: Implementing a plan that includes assessing vulnerabilities, controlling access, setting up alerts, and conducting regular audits.

The European Union has a unified research and innovation policy through its Food 2030 has with strategic initiatives to support enhance use of digital technologies, increased research and investment in innovation, and an assessment of socioeconomic impacts from moving towards digital platforms for the food system and food safety. Horizon Europe Sustainable Food Systems Partnership for People, Planet and Climate; EU Mission: A Soil Deal for Europe. The overall goals of Food 2030 is to support a transition to sustainable, healthy and inclusive foods in coordination with the European Green Deal farm to fork strategy and bioeconomy strategy and the belief that food systems have to change but that the socioeconomic impacts of data science and digital technologies on consumers and the broader agricultural and business community must be taken into consideration. SMART will be key to meeting these objectives and with the creation of new business models to support innovation, technology and new products, empower community change by fostering social innovation, and building a skilled workforce and consumer awareness.

Ecommerce and this new way of producing and delivering food is a concern in the EU with food safety recognizing that interactive platforms such as websites, mobile apps and social media are being used to educate consumers about food safety practices and provide consumers with a comprehensive product history including details of product origin, certifications such as organic, kosher or halal, handling and safety standards. Apps tied to consumer purchases can alert consumers in real time about recalls and safety notices associated with products they may have purchased. This fits in with the goal building a more flexible and agile financing framework, food chain measures will be integrated in other budget priorities such as research, innovation and digital policies across the member nations (Horizon Europe, Digital Europe).

The European market, more so than other markets is concerned with reducing food waste and waste from food packaging. Packaging with imbedded sensors for temperature, pH or volatile components, semipermeable packaging that controls the transfer of oxygen and water vapor and the composition of biodegradable food packaging can create new food safety challenges. Intelligent packaging that can monitor and then maintain optimal conditions to extend shelf life of perishable foods such as meat, seafood and fresh fruits and vegetables by adjusting temperature, gas composition, and humidity can extend food shelf life and reduce Smart packaging. Integrating traceability features into such packaging is relatively straightforward and coupled with IoT devices for inventory control this in turn will improve that can monitor and adjust temperature and humidity levels to ensure that produce remains fresh during transport and storage. Additionally, blockchain and IoT technologies enable better inventory management by providing real-time data on stock levels and product freshness. This reduces the likelihood of overstocking and spoilage, further enhancing sustainability. **Food safety in the future EU budget(2021-2027) https://food.ec.europa.eu/horizontal-topics/future-food-safety-budget-and-   policy/food-safety-future-eu-budget-2021-2027_en**

Conclusion

International markets recognize the importance of SMART food safety for managing production and food sales and its importance in food traceability. SMART provides the ability to both predict emerging food safety problems and to more effectively manage recalls if that become necessary. Implementation of data integration between regulatory agencies and businesses will be an important aspect of SMART food safety. Data integration combines data from multiple sources into a unified view making compliance-centric an evolving feature of food safety programs. SMART applications for food safety testing, quality assurance and production process controls will be where this technology is currently most quickly implemented in the food industry. The coupling of SMART with artificial intelligence platforms is providing the basis for more efficient product development through predictive models that can shortcut screening of ingredient, cultivar characteristics or microbial strains, product safety evaluation and clinical trials. Regulatory agencies are evolving new regulations and guidance to address the adoption of this rapidly changing food safety landscape.

Top of Form

Bottom of Form

Bottom of Form

ACKNOWLEDGEMENTS

This work is supported by Hatch-Multistate research funds, project accession no. 7006640, from the U.S. Department of Agriculture's National Institute of Food and Agriculture.

References

Akman. S., Berger, A, Bianchi, E., Primo Braga, C., Cristini, M. Dawar, K., Evenett, SJ, Helble, M., Kolev, G., Matthes, J., Mendez-Parra, M., Shumucker, C., Schwarzer, J., Tamura, A. and Xinquan, T. 2018. Mend it, don't end it: The case for upgrading the G20's pledge on protectionism. http:// thevoice.bse.eu/wp-content/uploads/06/protectionsim-20-t20-policy-brief-4th-draft.pdf. Accessed 7 October 2024.

Artificial Intelligence https://www.foodnavigator.com/Article/2024/09/16/AI-Unlocking-hidden-opportunities-for-food-and-beverage Accessed 11 November 2024.

Berkowitz, SA, Drake, C, Byhoff, E. 2023. Food insecurity and social policy: a comparative analysis of welfare state regimes in 19 countries. *International J Social Determinants of Health and Health Services.* 54(2):1177 https://doi.org/10.1177/27551938231219200

Brazil 2050: A vision for global food security - Atlantic Council Accessed 14 December 2024.

Broucke, C., van Pamel, E., Coillie, E.V., Herman, L and Royen, G.V. 2023.Cultured meat and challenges ahead: A review on nutritional, technofunctional and sensorial properties, safety and legislation. *Meat Science* 195:109006. https://doi.org/10.1016/j.meatsci.2022.109006

Bruun-Jensen, J. , Tanger, K. & Cascone, J. 2024. FSMA Section 204: Traceability and Tacking in the Food Industry. Using Digitalization to Track and Trace Foods in the Food Value Chain. Traceability and Tracking in the Food Industry | Deloitte US. Accessed 14 December 2024.

Charlebois, S., Latif, N. K., Ilahi, I., Sarker, B., Music, J. & Vezeau, J. 2024. Digital traceability in agri- food supply chains: a comparative analysis of IECD member countries. *Foods* 13(7), 1075. **https:// doi. org/10.3390/foods13071075**

China's Food Security: Key Challenges and Emerging Policy Responses (csis.org) Accessed 14 December 2024.

China: New and Amended National Food Safety Standards | USDA Foreign Agricultural Service Accessed 14 December 2024.

China Unveils National Food Safety Standard Legislation Plan for 2023 | ChemLinked Accessed 14 December 2024.

Diaz, R.E.B., Rojas, A.E. and Moncayo, C.M. 2023. Blockchain for Columbian organic coffee traceability on Hyperledger. Operations Research Forum DOI: https://doi.org/10.21203/ rs.3.rs-252513/v1 Accessed 10 December 2024.

EIT, 2021.European Institute for Innovation & Technology. Posted 18 July 2022. Accessed 7 October 2024.

EU Platform on Food Losses and Food Waste. This platform addresses food waste at every sage of the food supply chain involving both public and private sectors. Accessed 7 October 2024.

Food 2023: This research and innovation policy framework supports the transition towards sustainable, healthy, and inclusive food systems. It aligns with the European Green Deal, Farm to Fork strategy and Bioeconomy strategy. Accessed 7 October 2024.

Food Regulations – Singapore Statutes Online (agc.gov.sg). Accessed 7 October 2024.

Food Safety on the Future EU Budget (2021-2027): This initiative focuses on preventing food waste combating food fraud and supporting sustainable food production and consumption Accessed 7 October 2024.

Food safety in the future EU budget (2021-202) https://food.ec.europa.eu/horizontal-topics/future-food-budget-and-policy/food-safety-future-eu-budget-2021-2027 en Accessed 7 October 2024.

Food Safety and Security Bill in the works, will give more clarity on novel foods | The Straits Times Accessed 14 December 2024.

Food Standards Code | Food Standards Australia New Zealand Accessed 14 December 2024.

Future-forward food safety: Leading the way in India's next gen (foodinfotech.com) Accessed 14 December 2024.

GAO, 2023. Food prices:information on trends, factors, and federal roles. GAO-23-105846. March 28, 2023. GAO-23-105846, Food Prices: Information on Trends, Factors, and Federal Roles

GFI India for change: Gen Z demand overhaul of food system. (gfi-india.org) Accessed 14 December 2024

Halkier, B and Holm L. 2006 Shifting responsibilities for food safety in Europe: An introduction. *Appetite*. 47:127-133.

Health Canada 2022 to 2023 Departmental Sustainable Development Strategy Report - Canada.ca Accessed 7 October 2024.

Ideagen. 2024.Food fraud prevention:steps to keep your business safe-Idagen. https://www.ideagen. com/thought- leadership/blog/food-fraud-prevention-steps-to-keep-your -business-safe Accessed 7 October 2024.

Japan's Smart Food Value Chain: From Consumer to Agri-Food Industries. FFTC Agricultural Policy Platform(FFTC-AP)Accessed 14 December 2024

Jesus, A. 2022. Reimagining the food system European Environment Agency. Accessed 7 October 2024.

Longley, B. 2020. What Is Classical Liberalism? Definition and Examples ThoughtCo. Updated 29 June 2020. Accessed 9 September 2024.

Kim, R.B. 2009. Meeting Consumer Concerns for Food Safety in South Korea: The Importance of Food Safety and Ethics in a Globalizing Market J. *Agric Environ Ethics* 22:141–152 DOI 10.1007/s10806-008-9130-9 Korea. 2024.

Ministry of Food and Drug Safety>Our Works>Food>Food Safety/Ministry of Foood and Drug Safety (mfds.go.kr) Accessed 14 December 2024.
Next generation of food safety - Canadian Food Business Accessed 14 December 2024.

New Era of Smarter Food Safety, Blueprint. Modern Approaches for Moderns Times. New Era of Smarter Food Safety Blueprint | FDA. Content current as of 5 March 2024. Accessed 11 November 2024.

New Zealand Food Safety Strategy | NZ Government (mpi.govt.nz) Accessed 14 December 2024.

Post, M.J., Levenbert, S.,Kaplan, D.L., Genovesse,N., Fu, J.,Bryant,.J.,Megowitti,N.,Verzijden,K and Moutsatsou, P. 2020. Scientific, sustainability and regulatory challenges of cultured meat. *Nature Food* 1:403–415.

Ray, S., Sneha, K and Jangid. C. 2023. CRISPR-Cas9 for sustainable food production: impacts, recent advances and future perspectives. *Food and Humanity* 1:1458-1471.

Renewing the Food Policy for Canada: Towards 2030 – Food Secure Canada Accessed 14 December 2024.

#Safet2EatEU Campaign: This campaign educates citizens about food safety, including foodborne diseases, proper food handling techniques and the importance of reading food labels. Accessed 14 December 2024.

Science of safety: Singapore's Future Ready Food Safety Hub outlines novel foods, agriculture and aquaculture focus (foodnavigator-asia.com) Accessed 14 December 2024.

Shield, C. 2021. Who controls the world's food supply. Nature and Environment. Posted 8 April 2021 Who controls the world's food supply? – DW – 04/08/2021 Accessed 7 July 2024.

Shlomit, D., Tuckerman, A., Safina, D., Maor-Shoshani, A., Lavon, N. and Levenberg, S. 2023. Co-culture approaches for cultivated meat production Review Article Published: 12 June 2023 *Nature Reviews Bioengineering* 1: 817–831.

Singapore 2024. Smart food: novel foods, food security, and the Smart Nation in Singapore (researchgate.net) Accessed 14 December 2024.

TraceXTechnologies. 2024 Food Industry Trends – Future of Technology and Sustainability.

Ulea, A. 2023. Europeans are becoming more socially liberal, according to a new study | Euronews. Accessed 11 November 2024.

Van Zwanenberg, P and Millstone, E. 2005, The aftermath of 20 March 1996. In: BSE: Risk, Science and Governance. Oxford University Press., Oxford, UK Pp 199-208.

Varum, FJO, Hatton, GV and Basit, AW. 2013 Food, physiology and drug delivery. Int J Pharm. 457(2):446-460. doi: 10.1016/j.ijpharm.2013.04.034. Epub 2013 Apr 21.

Zedler, J.A.Z., Schirmacher, A.M., Russo, D.A., Hodgson, L, Gunderson, E., Matthes, A., Frank, S. Verkade, P. & Jensen, P. E. (2023). Self-assembly of nanofilaments in cyanobacteria for protein co-localization. *ACS Nano.* 17(24), 25279-25290. doi: 10.1021/acsnano.3c0860